NAKED LANDSCAPE

NAKED LANDSCAPE
JOHN SWANNELL

QUARTET BOOKS

LONDON NEW YORK

First published by Quartet Books Limited 1986
A member of the Namara Group
27/29 Goodge Street, London W1P 1FD

ISBN 0 7043 2609 4

Originated and printed in Great Britain by
S. W. Sharman & Co. Ltd, London

NAKED LANDSCAPE

3

7

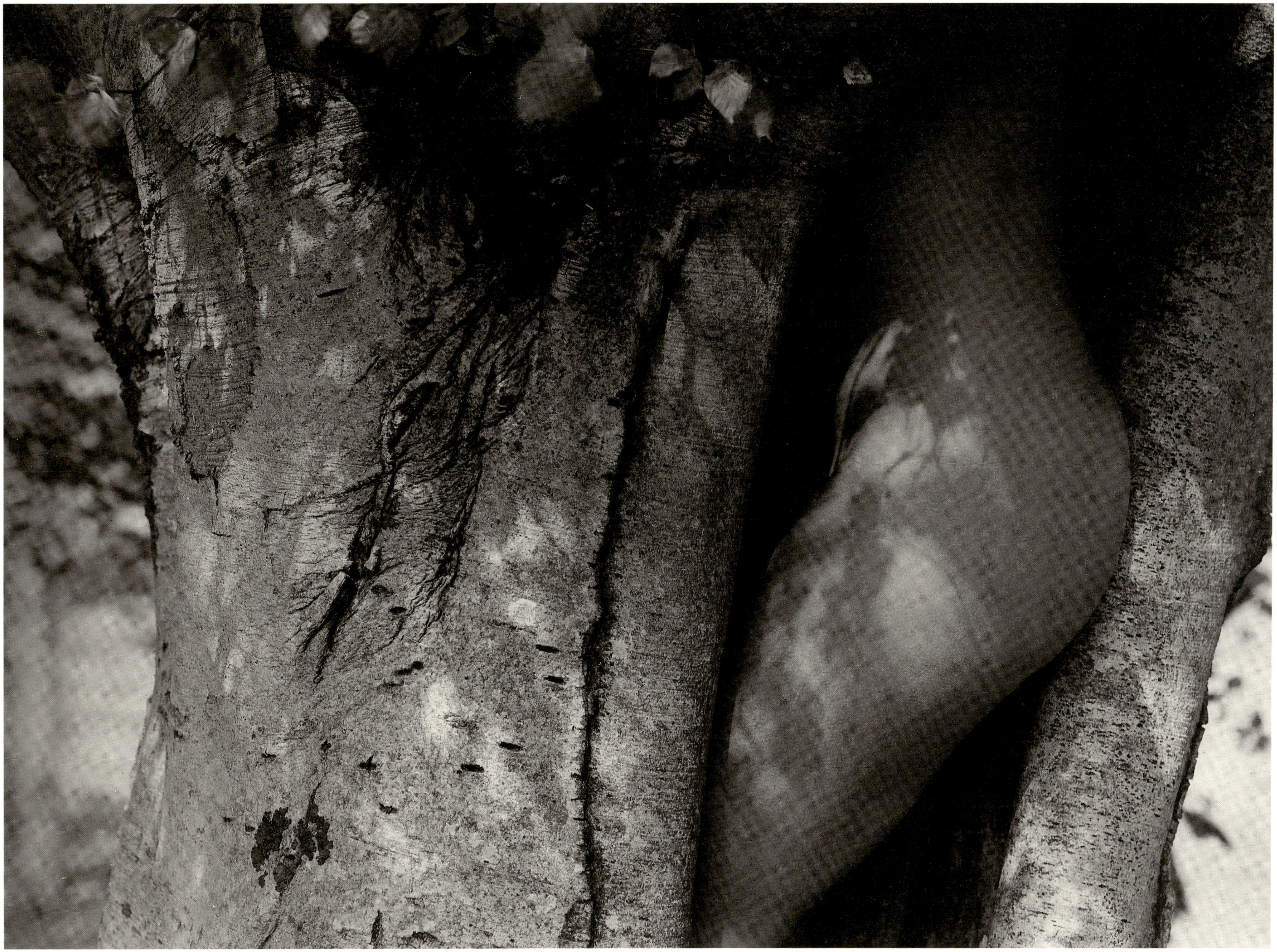

ACKNOWLEDGEMENTS

Mikaele
Sarah Knight
Becky Bain
Claire Adams
Terry Boxall
David Burnham
Michele
Highgate Gallery
Barry Taylor
Werner Teuffel
Michael & Frieda Yeo

And also very special thanks to Robin Bell and Martin Harrison.

and to my two assistants Karena Perronet-Miller and Robert Mackintosh.

Limited edition prints available at:
HAMILTON GALLERY, LONDON
and
PHOTOGRAPHERS GALLERY, LONDON